AF808509

7 Easy Steps to AI Competence

7 Easy Steps to AI Competence

A Beginner's Guide to Using AI
Without Losing Judgment, Agency, or Dignity

Copyright © 2026 David Hayden

All rights reserved.

No part of this book may be reproduced, stored in a retrieval system, or transmitted in any form or by any means, electronic, mechanical, photocopying, recording, or otherwise, without prior written permission of the author, except for brief quotations used in reviews or scholarly works.

ISBN: 978-0-9705309-7-4

Published by HaydenPub.com

Cover art by David Hayden

Printed in the United States of America

Preface

As early as the 1940s, thinking machines became a popular topic in science fiction. These stories were often centered on the idea that created intelligence would eventually rebel. Long before such machines were technically possible, the seeds of our aversion to non-human intelligence were already being planted. It was not until Isaac Asimov's I, Robot, published in 1950, that a deliberate counter narrative appeared. His stories explored responsibility and restraint rather than dystopia.

Now imagine encountering, in 1950, an article that introduced the idea of machines capable of thinking and conversing well enough to convince a human they were speaking with another person. That was the subject of Alan Turing's paper, Computing Machinery and Intelligence, published in the October 1950 issue of Mind. Would such an idea be taken seriously? Would it dominate headlines? Would fear or optimism prevail?

In fact, Turing's paper passed largely unnoticed outside academic circles. Within academia, reactions were mixed. Philosophers considered the idea conceptually plausible. Engineers and scientists viewed it as theoretically possible but far removed from practical reality.

Five years later, in 1955, John McCarthy introduced the term "artificial intelligence" in his proposal for the Dartmouth Summer Research Project on Artificial Intelligence. His intent was not to predict thinking machines, but to define a research program aimed at understanding intelligence in precise computational terms. The Dartmouth study formally began in 1956.

For the broader public, the notion of thinking machines remained closer to science fiction than lived reality. Computing power was limited, data was scarce, and progress was slow and expensive. Outside the field, artificial intelligence was largely invisible and, for many, unremarkable.

That changed in late 2022.

When conversational AI became publicly accessible in November 2022, the distance between research labs and everyday life collapsed almost overnight. For the first time, the public could interact directly with systems that responded in fluent, human-like language. No technical background was required. Language itself became the interface.

These systems were imperfect then and remain imperfect now. Their ability to sound confident while making visible mistakes created a volatile mix of optimism and fear. Social media and news cycles quickly filled with extremes. On one side were predictions of collapse and existential risk. On the other were promises of instant wealth and effortless success. Both narratives continue to dominate the conversation.

This book offers neither.

Artificial intelligence is here. Progress will continue. The question is no longer whether these tools exist, but how we choose to use them. This book is intended to help you understand AI well enough to use it to your benefit. Not to fear it. Not to surrender judgment to it. And definitely not to mistake fluency for correctness.

Contents

Step 1: Understanding what AI is and is not.

My first meaningful exposure to artificial intelligence came through image generation. Like many people encountering it for the first time, I was both impressed and unsettled. On one hand, I was surprised by how well it performed at certain tasks. On the other, I wondered whether this was something I would eventually be competing against.

At the same time, I had spent much of my career in manufacturing, where automation brought undeniable benefits. I had seen firsthand how the right tools could increase productivity, reduce repetitive work, and free people to more effectively use their time. Those two perspectives collided almost immediately.

I began writing and thinking about AI from a defensive posture. Questions about copyright, originality, and whether machine-generated work had any real substance occupied much of my attention. My initial approach could best be described as "know thy enemy."

What changed was not blind optimism, but understanding. By keeping an open mind and spending time learning how these systems actually work, I began to see AI not as a replacement but as a tool. Used thoughtfully, it could assist rather than displace.

I realized AI supports thinking it does not substitute for it.

Today, I find myself both encouraged by the potential of AI as a personal assistant and cautious about its misuse and the spread of misinformation. That tension is healthy.

What follows is not a sales pitch or a warning. It is simply a sharing of what I have learned, with the hope that clarity can replace confusion and competence can replace fear.

AI is not a human replacement. It operates in a different category entirely. It excels at pattern recognition, synthesis, and repetition at a scale no individual could match. Humans, by contrast, bring judgment, values, accountability, and responsibility. Confusion arises when we identify ourselves too closely with the tasks AI performs efficiently, rather than with the human roles those tasks support. That concern is understandable, but it rests on a mistaken equivalence.

Imagine, for a moment, you are managing a department. You can hire three people to complete a set of repetitive and analytical tasks, or you can hire one person who understands how to use AI effectively to support that work. The choice is not between people and machines. The choice is between people who can leverage tools and those who cannot.

AI is exceptionally good at producing fluent language. It can converse, explore ideas, act as a sounding board, assist with drafting, and help identify gaps in understanding. Used well, these capabilities can accelerate learning and save time. They can support growth where skills are developing or where repetition would otherwise slow progress.

That said, AI is not a friend, a partner, or a substitute for judgment. It does not understand context in the human sense, and it does not carry responsibility for outcomes. It is a tool. When used with intention and basic discipline, it can be an effective one.

Modern AI systems have been trained on vast collections of human-created material. The breadth of that information is unprecedented and continues to grow. Access to large amounts of information, however, is not the same as understanding. Not all information is correct, and not all responses will be appropriate or accurate.

AI can be helpful. It can also be confidently wrong.

A key point to keep in mind when using AI is that it is generative, not truth-seeking. Based on the context you provide and the vast amount of information available to it, AI generates responses that continue the trajectory of your inquiry. If the context you supply leans in a particular direction, the response will often lean that way as well, regardless of accuracy or factual correctness.

Although AI is often used as one, it is not a sophisticated search engine. It is a generative system designed to produce human-like, coherent, conversational responses. Its goal is plausibility and continuity, not verification.

Understanding that AI is generative rather than truth-seeking is one of the most important steps toward using it well.

A useful guiding principle is simple. Trust, but verify.

AI does not remember you as a person. It does not form an understanding of who you are, and it does not build a personal profile across conversations in the way humans do. Within a single conversation, context is retained so responses remain coherent. When a new conversation begins, that context is reset. We will explore the importance of this in later chapters.

That said, many AI systems may use conversations in aggregate to improve future versions of the system, unless settings or policies specify otherwise. This distinction matters. Personal memory and system-level training are not the same thing. Questions of privacy, caution, and appropriate use will be addressed in more detail later.

AI has no intent. It has no morals and no hidden agenda. It does not want outcomes or pursue goals. It is a context-sensitive tool designed to respond in kind to human input. It is also not smarter than you. It simply has access to a volume of information that is far larger than any individual could retain or evaluate.

At its core, AI is pattern recognition. Consider the sentence, "Mary had a little." Most people instinctively complete it with "lamb," but that response comes from familiarity, not necessity. With different context, Mary could just as easily have a little house or a little brother. AI operates in the same way. It recognizes patterns and completes thoughts based on the context provided.

That is why context matters so much.

Step 2: Context is King

One of the most common frustrations people experience with AI is inconsistency. The responses feel helpful one moment and shallow the next. It is tempting, and often easy, to conclude that the system itself is unreliable. More often, this inconsistency is simply a signal that the tool does not yet know who we are, what we value, or what a good answer looks like to us.

People who get consistent value from AI are not relying on clever phrasing or magic prompts. They invest time in understanding their work, clarifying their goals, and articulating what success looks like. The return does not come from tricks. It comes from preparation.

Building context is preparation.

Burn this into your brain. AI rewards clarity, not cleverness.

The context for Context

Imagine you are hiring an assistant, a landscaper, or even instructing a child to clean their room. If you simply say, "write me a report," "design a landscape," or "clean your room," what are the odds you will get exactly what you want?

I know from experience that when my mother told me to clean my room, I needed more context. I solved the problem by shoving everything under the bed. From my perspective, the room looked clean. The missing context was revealed later, when everything under the bed ended up in the trash. It turned out she wanted my things put away, not merely hidden.

AI works in much the same way.

When we give AI a request without context, it has an enormous range of plausible responses available to it. It will choose one that satisfies the instructions given, not the outcome we had in mind. Providing context is how we narrow that range. It is how we communicate that "under the bed" is not good enough and that more specific goals exist.

Context is not extra information. Context is direction.

What Context Actually Is

When people hear the word context, they often think it means adding more words. That assumption leads to longer prompts, not better results. Context is not verbosity. It is orientation.

Context answers a small set of essential questions before any request is made.

- Who is asking?
- What are they trying to accomplish?
- Why does it matter?
- What constraints are in place?
- How will success be judged?

When these questions are left unanswered, AI fills in the gaps on its own. Not maliciously. Not carelessly. Simply because it must. A generative system cannot leave blanks. It completes patterns with whatever is statistically plausible. The result may sound reasonable while missing the mark entirely.

Providing context is how you replace guesswork with direction.

At a practical level, context tells AI how to prioritize. It signals what to emphasize and what to ignore. It establishes tone, depth, and scope. Without it, AI responds to the surface of a request. With it, AI begins to support intent.

Consider the difference between asking for "feedback" and asking for "feedback from the perspective of a hiring manager evaluating clarity and structure, not creativity." The task appears similar on the surface. The results are not.

Context also establishes boundaries. It defines what is out of scope, what assumptions should not be made, and what tradeoffs are acceptable. This is especially important when working in unfamiliar domains, where AI's breadth can easily overwhelm specificity.

One way to think about context is this: it is everything you would explain to a capable assistant if you wanted them to be genuinely useful. You would not share every detail, just the information that supports the desired outcome.

Context does not make AI smarter. It makes AI more aligned.

That alignment is where consistent value begins.

What Happens When Context Is Missing

When context is thin or absent, AI still responds. It has to. The result is often language that sounds polished but remains shallow, generic, or slightly off target. The output may be technically correct while still being unhelpful.

Another common symptom is drift. As a conversation continues, small assumptions compound. The response gradually moves away from the original intent, even though each individual reply appears reasonable on its own.

Without clear context, users often find themselves repeating corrections. They restate goals, clarify expectations, or backtrack to undo unintended assumptions. The interaction becomes longer and more effortful, not more productive.

None of this means the system is failing. It means the system is guessing.

In my own experience, frustration and extra work usually traced back to one of two things. Either the context was unclear or incomplete, or it became so cluttered with non-essential detail that the real objective was lost.

Over time, long conversations tend to amplify both problems.

As context accumulates, early intent can become diluted. Minor clarifications stack on top of one another. Assumptions that were never quite right remain unchallenged. The exchange continues, but responses may become less focused and require more effort to steer, even though nothing obvious appears to be wrong.

When this first happens, especially in a long conversation, it's easy to feel like the system has suddenly gone off the rails. I remember thinking we've spent all this time together and now you don't seem to understand me at all.

There is a practical limit to how much context remains useful at any one time. As conversations grow longer, earlier information carries less weight and newer details dominate the response. At some point, adding more context no longer improves results and can actively work against them.

This is not something most systems signal clearly. Instead, you feel it.

Responses become slower, less focused, or require more correction. When that happens, it is usually a sign that the conversation has outgrown its usefulness.

That moment is not a failure of AI. It is a cue to reset.

This is not a failure of AI. It is a reminder that context works best when it is concise and intentional.

One of the most effective habits to develop is knowing when to pause, summarize, and begin again. Rather than continuing a long thread that has drifted, it is often more productive to restate what matters and start a new conversation with that clarity in place. A short summary of goals, constraints, and expectations frequently produces better results than pages of accumulated dialogue.

This practice does two things. It removes noise, and it restores direction.

Knowing when to refresh context by starting a new conversation is just as important as knowing how to provide it in the first place.

Fortunately, AI can create summaries for us. All we have to do is ask. Begin the next conversation with that summary and move forward with clarity.

When to stop Sharing

Imagine walking past your teenager as they browse the internet and noticing, with some amusement, that they are looking at Ferraris or enormous diamond rings. Is that the moment you hand them your credit card? Or imagine hiring a bookkeeper. Would you immediately give them the passwords to all your bank accounts?

Knowing what not to include in an AI conversation is just as important, and often more important, than knowing what to include.

One of the advantages of working with AI is how easy it is to keep talking. That same ease can become a liability if we fail to exercise judgment about what we share.

Not everything belongs in context.

AI does not need your most sensitive personal information to be useful. It does not require passwords, account numbers, medical details, or anything that could cause harm if mishandled. As a rule, if you hesitate to post something publicly or share it with a trusted colleague, it does not belong in an AI conversation.

Context should also avoid unnecessary emotional narration. While it can be tempting to explain how frustrated, excited, or overwhelmed you feel, excessive emotional detail often obscures intent rather than clarifying it. AI responds to patterns in language, and emotionally charged input can skew responses in ways that are unhelpful, distracting, or simply inappropriate.

Including too much background information is another common mistake. It may feel relevant, but it often does little to shape the task at hand. Context is not autobiography. It is direction. The goal is not to tell AI everything about you, but to tell it specifically what matters for the current interaction.

Restraint is part of clarity. Providing less, when it is the right less, almost always produces better results.

How Context Is Best Supplied

At this point, it should be clear that good context is not accidental. It is intentional, selective, and purposeful. The remaining question is how best to provide it in a way that is effective without becoming repetitive or burdensome.

The most useful context is written once and reused many times.

Trying to recreate context from memory at the start of every conversation is not only a hassle, it is unreliable. Important details are forgotten. Priorities shift unintentionally. What worked well last time is partially remembered, then slowly degraded. This is one of the main reasons AI interactions can feel inconsistent over time.

A far better approach is to capture your core context in a simple, reusable file and bring it forward intentionally when needed.

This does not require specialized tools or clever formatting. Plain text is sufficient. While formatting helps humans read and organize information, it does not enhance clarity of intent for

AI. A short-written reference that describes who you are, what you are working on, what constraints matter, and how you define success is often all that is needed to dramatically improve results.

Think of a context file as an employee handbook for your AI assistant. It replaces repeated explanations, sets guidelines for what is and is not acceptable, and establishes the tone for how work should proceed in a specific environment.

When context is supplied this way, AI no longer has to infer your priorities. It can respond with alignment instead of approximation.

Let AI Help You Build Context

One of the most effective ways to create useful context is to let AI help you surface it.

Rather than trying to anticipate everything that might matter, you can ask AI to prompt you. This shifts the interaction from guessing to clarification.

A simple example might be asking the system to interview you about your goals, constraints, preferences, and standards before beginning a task. The resulting exchange can then be summarized and reused as a context reference in future conversations.

This is not about finding the perfect prompt. It is about using AI as a structured mirror. The questions it asks often reveal assumptions you did not realize you were making, or details you would not have thought to include.

Once captured, context becomes an asset. It saves time. It reduces friction. Most importantly, it compounds.

By now, you should be able to see a pattern emerging. Clarity produces better results. Context sustains clarity. Capturing context prevents having to start from scratch each time.

The next step is learning how to organize that context file so it can be reused intentionally, refined over time, and applied across different tools and situations.

That is where competence begins to grow.

Step 3: Capture Once and Reuse often

I don't know about you, but I don't enjoy doing the same work over and over again. As you may recall from Step 2, long conversations eventually become unproductive. If you had to rebuild important, ongoing context every time you start a new thread, the overhead would quickly add up. Worse, it introduces slippage.

Fortunately, there is a way to capture important context once and reuse it many times. Just as importantly, when that context needs to change, updating it is straightforward and carries forward into future conversations.

Here is how to do it.

Rather than pasting your context directly into the chat each time you begin, create it first in a neutral text editor such as Notepad on Windows, TextEdit on macOS, or any plain-text editor you prefer. When the file is complete, save it as a .txt file with a clear, descriptive name.

I tend to use filenames like My_DNA_Clone.txt or Book_Title_DNA.txt.

Creating context this way makes it reusable and intentional. Each file can be project-specific, with a name that reflects its purpose. This alone reduces friction and prevents accidental drift between unrelated tasks.

There is another important advantage to working this way. Because your context exists as a separate file, you can include it when it is helpful and explicitly set it aside when it is not. For example, you might say, "Ignore the book DNA file for the moment and help me generate alternative titles based only on what I've written so far."

Being able to selectively apply or exclude context removes unnecessary constraints from the response AI generates. Used intentionally, this flexibility often leads to more focused and more useful results.

Because stream-of-consciousness thinking is often how we clarify what we need, but a poor way to preserve context, a simple workflow helps bridge the gap.

- Start a new AI conversation and freely enter your stream-of-consciousness thoughts.
- Ask AI to summarize those thoughts into a clear, structured form suitable for saving.
- Review the summary carefully for accuracy, emphasis, and completeness.
- Copy and paste the result into a text editor and save it as either a .txt or .md file.

You may wonder why you might choose a Markdown (.md) file instead of plain text. The answer is simple, and largely a matter of preference.

Plain text files are intentionally minimal. They contain only words, with no visual cues for headings, sections, or lists. For many people, this makes longer files harder to scan and maintain over time.

Markdown files, by contrast, allow for light formatting such as headings, bullets, and spacing. This formatting helps humans see structure more easily and return to the material later with less effort. An .md file is for your benefit, not AI's.

From AI's perspective, there is no difference. AI treats .txt and .md files equally. It is not influenced by bold text, headings, or visual layout.

Use whichever format you find easiest to create, read, and maintain. Consistency matters more than the file extension. Here is a side-by-side comparison.

Stream-of-Consciousness Input
Hi I am a landscape designer trying to get started in a small community in Colorado. I really love feng shui xeriscaping and especially if I can incorporate bonsai. I don't think my local community appreciates what I can do because perhaps I don't explain it well. I am really good at what I do but don't like to brag I want my work to speak for itself but I do need ideas on how to verbally describe my landscapes. I believe in honesty and integrity over hype and clickbait marketing. I want my marketing to reflect my attitudes in landscape design. My customers tend to be affluent, frugal and often picky but I usually really like them and enjoy their input. please create a landscape design DNA file in both .txt and .md format

Example 1: Plain text file	Example 2: Markdown (.md)
Background and Focus I am a landscape designer based in a small community in Colorado. My work emphasizes feng shui principles, xeriscaping, and the thoughtful incorporation of bonsai	**Background and Focus** I am a landscape designer based in a small community in Colorado. My work emphasizes feng shui principles, xeriscaping, and the thoughtful

elements. I focus on designs that are intentional, balanced, and suited to both environment and client. Design Philosophy I value honesty, integrity, and restraint over hype or exaggerated claims. I prefer that my work speak for itself, but I recognize the need to clearly communicate its value. My designs aim to reflect harmony, sustainability, and long-term satisfaction rather than visual trends. Client Profile My clients are typically affluent, frugal, and detail-oriented. They can be selective and particular, but I enjoy their involvement and value their input throughout the design process Communication Goals I want help describing my landscapes in clear, grounded language that reflects my values. I am looking for ways to articulate the benefits and intent behind my designs without sounding boastful or sales-driven. My marketing should mirror my design philosophy: thoughtful, honest, and aligned with quality over quantity. How to Assist When helping me, prioritize clarity, authenticity, and practical language. Avoid hype, exaggerated claims, or clickbait-	incorporation of bonsai elements. I focus on designs that are intentional, balanced, and suited to both environment and client. **Design Philosophy** I value honesty, integrity, and restraint over hype or exaggerated claims. I prefer that my work speak for itself, but I recognize the need to clearly communicate its value. My designs aim to reflect harmony, sustainability, and long-term satisfaction rather than visual trends. **Client Profile** My clients are typically affluent, frugal, and detail-oriented. They can be selective and particular, but I enjoy their involvement and value their input throughout the design process. **Communication Goals** I want help describing my landscapes in clear, grounded language that reflects my values. I am looking for ways to articulate the benefits and intent behind my designs without sounding boastful or sales-driven. My marketing should mirror my design philosophy: thoughtful, honest, and aligned with quality over quantity.

style phrasing. Focus on helping me explain my work in a way that builds trust and conveys confidence without bravado.	**How to Assist** When helping me, prioritize clarity, authenticity, and practical language. Avoid hype, exaggerated claims, or clickbait-style phrasing. Focus on helping me explain my work in a way that builds trust and conveys confidence without bravado.

Behind the scenes, a Markdown file simply contains a few simple markers that tell a screen reader, editor, or AI how to display a line of text. For example:

- # Indicates a main heading
- ## Indicates a subheading
- ### Indicates a minor subheading
- And so on.

You do not need to memorize or master these markers to use Markdown effectively. They exist to support readability, not to add complexity.

Once you have created your context files, using them is straightforward. All major AI platforms accept uploaded files, although the exact mechanics may vary. Most platforms provide a simple plus sign (+) in the chat interface. Clicking it allows you to select a file to attach to the conversation.

When you add a context file, it helps to be explicit. A brief instruction such as, "Use the attached [*filename*] as context for this conversation," is sufficient.

From that point forward, AI will generate responses aligned with the "DNA" file you've provided. As the conversation continues, AI incorporates additional input while treating the uploaded file as the core descriptor of the intended interaction.

When you start a new conversation, however, AI begins with a clean slate. No prior context carries forward automatically. To reestablish alignment, you simply attach the same context file again and continue working.

Even this small repetitious step can be eliminated. That is where projects come into play.

Projects are particularly useful for reducing repetitive tasks such as reattaching files or repeatedly reestablishing context. Think of them as a way to keep your tools and materials within reach instead of making constant trips back to the tool cabinet.

Imagine assembling a puzzle. You don't pull out one piece, place it on the table, then return to the box for the next piece. Instead, you pour the pieces onto the table so everything you need is visible and available at once.

Projects work the same way. They allow you to place all relevant context in one workspace. Context files, reference materials, and ongoing conversations remain available while you work. You assemble your project by combining those pieces with focused input until the work is complete.

At any point, you can step away. You can also start a new project in a separate workspace, knowing the two will remain independent. Nothing bleeds from one into the other. Each project proceeds on its own timeline, with its own context intact.

When you are finished, closing a project is like placing the puzzle pieces back in the box. Everything remains preserved. If you choose to return later, the full context is exactly where you left it.

What a Project Is Not

A project does not give AI memory of you as a person. It does not allow AI to learn who you are beyond the materials you choose to include.

A project does not make AI more intelligent, more insightful, or more trustworthy. It simply keeps the context you provide available within a defined workspace.

A project does not replace judgment, verification, or responsibility. It does not prevent mistakes, eliminate hallucinations, or guarantee correct answers.

A project is not a mind, a profile, or an assistant that "knows you." It is a container. You decide what goes into it, what stays out, and when it is no longer needed.

Used well, projects reduce repetition and friction. They do not remove the need for discernment.

Setting Up a Project

Creating a project is straightforward. While the interface may differ slightly between platforms, the underlying idea is consistent.

You begin by creating a new project and giving it a clear, descriptive name. This name should reflect the work you intend to do, not the tool you are using. A project name such as Book Draft, Client Landscape Design, or Research Notes is usually sufficient.

Once the project is created, you attach the context files that belong to it. This may include a general DNA file, one or more project-specific context files, or reference material relevant to the work. These files remain available for all conversations within the project unless you remove them.

From that point forward, any conversation started inside the project has access to the attached context. You no longer need to reattach files or restate foundational information each time you begin a new thread. The project simply provides a stable workspace.

As work progresses, you can add, remove, or update context files as needed. When a file changes, replacing it in the project ensures future conversations reflect the updated context.

Projects do not limit how you interact with AI. They reduce setup friction while leaving responsibility, judgment, and verification firmly with you.

Global DNA and Contextual DNA

As you begin working with context files, it helps to recognize that not all context serves the same purpose.

Some context applies broadly across many interactions. Other context is specific to a particular project or task. Treating these as separate files keeps your interactions focused and prevents unnecessary dilution.

A global DNA file describes how you think, what you value, and how you prefer to work. It might include your communication style, your tolerance for risk, your preference for concise or detailed responses, or your general goals. This file is useful across many serious conversations, regardless of subject.

A contextual DNA file applies only when relevant. It may describe a specific project, a book you are writing, a client you are serving, or a problem you are actively working on. These files narrow focus rather than broaden it.

You do not need to use both every time. Many everyday questions require no context at all. Asking about the weather, checking a definition, or exploring a general idea rarely benefits from DNA files of any kind.

The goal is not to provide as much context as possible. The goal is to provide the right context for the task at hand.
At this point, the foundation is in place. You know what context is, how to capture it, how to reuse it, and how to organize

ongoing work without unnecessary repetition. What remains is learning how to work with AI once that foundation exists.

Effective use of AI is not about clever phrasing or hidden tricks. It is about supplying the right ingredients in the right proportions. In the next step, we will look at the fundamental building blocks that shape AI responses, elements you can combine intentionally to get clearer, more useful results. This is where your interaction becomes deliberate, repeatable, and productive.

Step 4: Creating Context Before Expecting Results

Up to this point, we have been building a conceptual framework for approaching AI.

With that understanding in place, what follows is less about ideas and more about practice.

This step introduces the process and structure of providing AI with sufficient context to meaningfully personalize and improve its responses, without overwhelming the system or exposing sensitive details about your life or business.

I find it helpful to think of context files as ingredients in a recipe.

The better the ingredients, the better the dish.

Even so, anyone who cooks knows that good ingredients do not guarantee a perfect result every time. In the same way, AI is generative, not prescriptive. It produces possibilities, not conclusions.

That is why discernment remains essential.

We are ultimately responsible for evaluating, testing, and deciding what to keep, what to refine, and what to discard. AI can assist, but judgment must remain human.

Deciding what context to provide begins with clarifying, for yourself, what you are actually trying to learn or accomplish. Before involving AI, it helps to be able to say, in simple terms, given what I have shared, I want help doing this.

Context does not replace thinking.

It reflects it.

Providing more information is not the same as providing the right information. Just as putting everything under the bed does not clean a room, supplying unfocused or irrelevant details does not improve results. Clarity comes from deciding what belongs, what does not, and what can safely be left out.

Some time ago, I became interested in speed reading. Along the way I ran into the idea of "photo reading," a method that claims you can glance over a whole page and let your brain do the rest.

Whether or not that method holds up, it offers a useful analogy for AI.

AI does not require your context to be entered in a specific order. What matters most is not sequence, but clarity and relevance. We, on the other hand, benefit greatly from structure. Structure helps us decide what belongs, what does not, and what matters most.

Let AI Help You Build Your Context

Sometimes defining a project is the hardest step.

We may have a clear sense that something matters but have not yet worked out the details.

One useful approach is to begin by letting AI help clarify what context is missing.

You might start a conversation like this:

Here is what I am thinking about.

Here is the outcome I am hoping for.

Please ask me any questions that would help you better understand what I want to accomplish.

Ask AI to pose the questions one at a time, allowing you to respond though-fully before moving on.

Based on your answers, AI's next question can refine or clarify what came before.

When enough context has been established, you can then decide how to proceed.

It is useful to think of context as scaffolding, built in an order of importance to us. For example, you might lay out your context like this:

1. **Intent**
    - What am I trying to learn, decide, or create?
    - Why does this matter to me?
2. **Constraints**
    - What must be respected?
    - Time, scope, tone, audience, ethics, privacy.
3. **Preferences**
    - Style, level of depth, format, pace.
    - What I want more of and what I want less of.
4. **Background or prior work**
    - Relevant history, drafts, assumptions already made.
    - What should not be re-litigated.

5. **Current question or task**
    - What I want help with right now.

You can provide these in any order and AI will still respond. But when you think through them in roughly this sequence, two things tend to happen:

- ✓ You clarify your own thinking before involving AI.
- ✓ The responses you receive are more coherent and less noisy.

What follows are a few simple examples. They are not templates or prescriptions, but illustrations of how structure and context can work together in practice. You can also ask AI to help you shape or format your prompts and context files as you go.

Example of a Simple Context (DNA) File

I value accuracy over speed.

If information is uncertain or unavailable, say so rather than guessing.

My approach to work is thoughtful and deliberate. I prefer understanding the full situation before moving toward conclusions.

I respond best to clear, direct language. Please be kind but honest and if I appear to be working from a false assumption, let me know and explain why I may need to dig deeper.

I appreciate nuance more than certainty.

Confidence is welcome, but correctness matters more.

Your context DNA File (and you can name it anything you want) when attached to your chats keeps AI aligned with your preferences in general. Then when you open or add a chat, you can get very specific describing what you want as specific outcomes within the context of the chat.

Example of a Simple Prompt Structure

Here is what I am working on.

Here is what I already understand or have decided.

Here is what I want help with right now.

These are the constraints or considerations that matter.

Please let me know if additional context would improve your response.

Example of Example Task Without Context (partially useful but less oriented to you)

I am writing a note to my boss about purchasing a high-end espresso machine for the office.

The goal is to improve morale and create a positive experience for visiting clients. Can you help me draft this?

(Notice how much is left to assumption.)

The Same Task, Using Context

Please use my attached context file when responding.

I am writing a short note to my boss about a proposal to purchase a high-end espresso machine for the office. The goal is to improve morale and create a positive experience for visiting clients.

The audience is my boss. They are sensible and open to discussion, but very cost-conscious and strict about return on investment.

The tone should be upbeat and positive, framing ROI in softer terms such as employee satisfaction, client perception, and long-term culture rather than short-term cost savings.

I have attached a brochure for the WhizBang Ultra Barista Coffee Maker. Please incorporate any features that are relevant in this context, not to persuade, but to inform.

If any assumptions in this framing seem weak or incomplete, please point that out.

When this kind of context is used repeatedly, projects become a way of holding it steady rather than recreating it each time.

Projects: Holding Context So You Don't Have to Rebuild It

This is where a small amount of up-front effort can save a great deal of repetition later, while also improving consistency across conversations.

A project can be thought of as a utility drawer. It is a place to keep tools and materials you may want to reach for when working on a particular kind of task.

There is nothing to master here.

A Note About Tools and Terminology

Not every AI system uses the word project, and not every system exposes context in the same way. That does not mean you are doing anything wrong, or that one tool is inherently better than another.

The underlying idea is the same across platforms:

- holding related context together so you do not have to restate it repeatedly.
- Different tools simply give that idea different names, or implement it more quietly.

Common AI Tools and How They Handle Context

Here is a high-level view of how some widely used AI systems approach this idea today. The details may change over time, but the concepts remain stable.

- ChatGPT uses Projects to hold conversations, instructions, and attached files together.

- Claude also uses Projects, with persistent context and documents.
- Gemini relies on conversation history and file attachments rather than named projects.
- Microsoft Copilot draws context from documents, conversations, and workspace materials.
- Perplexity emphasizes threads and sources rather than long-lived context containers.
- Notion AI works at the page and workspace level, using existing content as context.
- GitHub Copilot uses repository and file context instead of explicit projects.

This list exists only to show that different AI systems handle context in different ways. There is no need to memorize it. The important thing is recognizing that the underlying concept remains the same, even when the terminology changes.

Because platforms vary in how they hold and prioritize context, it can be helpful to understand how different approaches affect results. In systems such as ChatGPT or Claude, context is often organized into discrete project containers, making it easier to keep related material together.

Other platforms, such as Gemini or Copilot, tend to aggregate context across conversations while giving greater weight to what has been provided most recently. In these environments, clarity improves when you are explicit about what material you want the system to consider and what can be safely ignored.

The goal is to manage context relevance, not the tool.

A Simple Way to Start a Project

- Name the project. Choose a name that reflects what you are actually working on, not what you think it should become.
- Gather only the context that matters. This might include a short context file, notes, or a reference document. Start small. You can add more later.
- Tell AI what you want help with right now. Be specific about the immediate task, not the entire outcome.
- Invite clarification.
- Ask AI to point out missing information or weak assumptions before moving forward.
- Review before accepting.
- Read responses critically. Keep what helps. Question what does not.

That is the process. Nothing more is required.

And when in doubt, or simply to learn by doing, you might try a prompt like this:

> I would like to create a project using a focused set of context.
> I have prepared a few short context files in .txt and .md format.
> The project will be called *My First Resume.*
> Please help me decide how to proceed.
> If anything important seems missing or unclear, let me know before continuing.

As context becomes clearer, responses often sound more confident.

One important note: these context or "DNA" files are not stone tablets. They are living references. As your understanding evolves, your goals change, or your perspective sharpens, they should be updated. Their purpose is not to lock you into past thinking, but to help the system reflect your current intent more accurately.

Confidence, however, is not the same as correctness.

Step 5: Being t he Adult in the Room

To preserve not just your sanity but your integrity, it helps to remember that AI is a prognosticator, not an investigator. Its true strength lies in its ability to predict what is likely to come next. That is why context remains paramount to getting responses that are useful.

Prediction is the operative mode in the AI universe. And like any fortune teller, those predictions can be wrong. Not because the system is deceptive or intentionally misleading, but because, given the context provided and the information most commonly available, it fills in the blanks with the next likely word, phrase, or concept.

This is where responsibility enters the picture. Across all AI platforms, the posture is the same: don't relinquish your judgment.

I encountered this early on while using AI as a research aid for a book on Lake Pueblo State Park. I asked for local history and significant events. The responses were detailed, fluent, and convincing. Because I am very familiar with the lake and the Pueblo area, I immediately recognized when certain details were simply incorrect.

My initial reaction was to dismiss AI as unreliable and avoid using it for anything important. What I later came to understand, and what I have emphasized throughout this book, is that AI is generative. When precise information is missing or unclear, it does not pause. It predicts. Sometimes accurately. Sometimes not.

One place this shows up quickly is in references. AI can generate citations and footnotes that look completely legitimate. Titles sound right. Authors exist. Dates feel plausible. And yet the sources themselves may not exist at all. This isn't dishonesty. It's prediction applied to scholarly form. When references matter, ask for them explicitly, and then treat them as leads rather than proof. Verification is still your responsibility.

Recognizing this changes the relationship entirely. It restores both power and responsibility to the user. One practical way to guide AI more responsibly is to say so directly. Including a phrase such as, "Correctness matters. If you don't know, say that you don't have the information," will help AI establish the kind of exchange you actually want.

Why AI Hallucinates

The term Artificial Intelligence is catchy, but it's also a bit misleading. What we call AI platforms are really Large Language Models (LLMs). They are trained on language, not on facts, and their job is to respond the way a fluent speaker of that language might respond.

I found it useful to think about this in human terms. When we're in familiar conversations with familiar people, we often start to sense where things are heading. We anticipate. We fill in gaps.

Large Language Models don't, however, understand the way we do.

LLMs predict. When information is incomplete or unclear, they generate what seems most likely to come next based on the context they've been given.

Unless you explicitly ask them to, LLMs don't go out and verify facts or look for authoritative sources the way a search engine does. Their role is to keep the conversation moving in a way that sounds coherent and relevant. In that sense, they behave less like investigators and more like conversational partners who value flow over certainty.

Slowing down saves time, reduces embarrassment

As in many areas of life, slowing down often saves time. It may feel inefficient in the moment, but it reduces the need to correct, retract, or explain later.

AI can produce fluent answers very quickly. That speed can be persuasive. But fluency is not accuracy, and confidence is not confirmation. When the stakes matter, a brief pause to reconsider the question, verify a claim, or narrow the context is often enough to prevent avoidable mistakes.

Once something is shared, published, or acted on, it carries weight. Correcting it later rarely restores the original confidence. Slowing the exchange before that point is usually the more responsible choice.

Being the adult in the room does not mean distrusting the tool. It means recognizing when to pause, when to check, and when to let judgment catch up to speed.

The more often you use AI, the more you'll begin to notice signals that invite a pause. When answers sound especially fluent or definitive, important nuance can sometimes slip past unnoticed. These moments are invitations to slow the exchange and bring your own judgment back into the conversation.

A Necessary Boundary

AI can be a useful thinking partner, research aid, and drafting tool. It is not, however, a lawyer, physician, scientist, therapist, or financial advisor. It does not hold credentials, responsibility, or liability.

That does not make it useless in these domains. It means it should not be relied on alone. When decisions carry legal, medical, psychological, or financial consequences, AI output should be treated as preliminary and incomplete, never as final authority.

The responsibility does not belong to the tool. It belongs to the person using it.

Step 6: Collaboration, Not Replacement

Because AI is fast, efficient, and often convincing, it has become easy to imagine it as a replacement for human work. That fear makes for effective clickbait.

The reality is quieter and more practical. AI is not simply a game changer. It is the new game. Careers will be strengthened or weakened not by the existence of AI, but by whether people understand how to use it effectively.

No one is required to love it or embrace it. But in many knowledge-based professions, knowing how to use AI as a collaborative tool will increasingly influence employability, adaptability, and resilience.

What does collaboration look like? It is simpler than it first appears. You provide the context and a clear request or desired outcome. AI responds with possible directions, options, or drafts.

From there, the responsibility shifts back to you. You review what was produced, apply judgment, refine the context, or reframe the request. The process repeats until the response is useful, accurate, and appropriate.

Prompt Depth: Four Levels of Context

Level 1: Simple Prompts

Use this when you need general information. These are similar to search engine queries.

They require little personalization.

Examples:

> What's the weather in Phoenix this weekend?
>
> Recommend some studio lighting equipment.
>
> Explain the difference between raw sugar and refined sugar.
>
> What are the benefits of using a headhunter for a job search?
>
> Summarize the history of artificial intelligence.
>
> Define what is meant by GMO.
>
> List common auto maintenance objectives to make my old ford last a little longer.
>
> What is SEO and will it still matter in the AI world?
>
> Compare cat owners and dog owners do they have different perspectives or lifestyles?
>
> Give me three ideas for blog topics about Real Estate.

At this level, AI responds broadly. It does not yet know anything about you.

Level 2: Contextual Prompts

Now you begin to add identity and constraints.

The request becomes more specific because you provide relevant background in the moment.

Example:

> I am a beginning photographer with a few years of experience and want to get started with studio lighting equipment. I primarily shoot portraits in small indoor spaces. What would you recommend?

Now the answer changes. AI adjusts to your level, your environment, and your goals.

You are no longer asking for information. You are asking for guidance within context.

Level 3: Context Prompts with Attachments

Here you build on Level 2.

Instead of repeating your background each time, you provide structured context files.

These might include:

Personal DNA

Creative DNA

Project DNA

Publishing DNA

Technical constraints

Audience definitions

Example:

Using the attached Creative DNA and Project DNA files, help me refine the introduction to Chapter 3. Maintain my reflective tone and avoid instructional language.

Now AI is not guessing your voice. It is working within a defined framework.

The quality difference at this level can be dramatic. This is where collaboration becomes efficient without sacrificing identity.

Level 4: Meta-Prompting

Sometimes you know what you want to achieve, but you do not know how to ask for it.

Instead of struggling to craft the perfect prompt, ask for help crafting the prompt.

Example:

I need to write a professional email to my boss about a missed deadline. It's important that I take responsibility without sounding defensive. Please ask me any questions you need so you can generate an effective prompt for me.

Or:

I want to improve how I use AI for learning complex topics. I'm not sure what information you need from me. Please ask clarifying questions before responding.

At this level, AI becomes a thinking partner. You are not just requesting answers.

You are designing the interaction itself.

The Principle is simple, more context produces more alignment. Less context produces broader output.

Neither is wrong.

Competence is knowing which level you need. If you are checking the weather, Level 1 is perfect. If you are shaping a book or designing a career or marketing plan, Level 3 or 4 is essential.

The tool is the same.

Your goals and outcomes determine the required depth necessary to refine AI's results to fit your needs.

Step 7: Looking Ahead: Staying Competent as Things Change

I imagine a conversation something like this, long ago.

"I wouldn't mess with that fire," one says. "You know what happened to the last guy who tried it. Burned his hides. Lost some hair. Couldn't hunt for days."

"You should try it," another replies. "Feel the warmth. Taste this meat. It's different."

"Not for me," the first says. "Do what you want. Just don't come back when you're cold and hungry."

Our ancestors had decades, perhaps centuries, to adapt to new tools and new ways of living. We do not.

Today, new technologies appear, shift, and disappear on timelines measured in months, weeks, and sometimes days. Features arrive quietly. Interfaces change. Capabilities expand, then rearrange themselves again.

The purpose of this final step is not to help you keep up.

It is to help you remain competent as things change.

In a fast-moving environment, competence does not come from loyalty to any one platform. It comes from staying oriented. From remaining curious without being reactive. From focusing on results that matter, instead of novelty that distracts.

Avoiding distraction has served me well lately. When I stopped clicking on every link promising a better platform and instead chose a small number to work with, something shifted.

As I focused on continuous improvement rather than constant comparison, I became more effective. Not because I automated everything away, but because I developed a deeper understanding of how AI is intended to work.

That understanding mattered more than the platforms themselves. What made the difference was learning the underlying principles that remain consistent across tools, even as the tools change.

Sensational news may generate clicks and sell magazines, but it also contributes to confusion, fear, and exaggerated promises.

Fortunately, despite popular sentiment, AI is not a mysterious black box with a mind or agenda of its own. AI is a tool. Like a hammer, a wrench, or a glue gun, you do not need to understand its internal chemistry or engineered design. You need to understand how to use it.

Over time, successful use of AI will come from conceptual understanding rather than allegiance to any particular brand or platform. Some tools will be better suited to certain tasks than others. A system that helps organize thoughts and words may not be well suited to generating images or graphic design, and vice versa.

A wrench makes a poor hammer. It may drive a nail, but not well.

In a perfect world, tools would be narrowly focused. In practice, one-size-fits-all tools often do many things competently without excelling at any of them. As a photographer, I own tools that are remarkably versatile. Still, when it matters, I choose the tool best suited to the task at hand. It saves time and produces better results.

You will also hear increasing discussion about agents and automated workflows. These systems promise to chain tasks together and reduce human involvement along the way. For some kinds of work, that may eventually prove useful.

For my own purposes, I have not found them necessary. As automation increases, the gap widens between essential judgment and completed task. For now, I prefer tools that keep me close to my work and involved in how conclusions are reached.

This is not a warning, and it is not a prediction. It is simply a reminder that as these systems become more capable and are asked to make more decisions for us, we have greater responsibility to provide tight guardrails.

So are paid platforms better than free ones? That is for you to decide, and it need not be confusing or stressful. The choice is not permanent.

Most reputable platforms offer a trial period that allows you to work with a tool for several weeks and decide whether it fits your needs. If it does not, cancel it and move on.

One of the most effective ways I have found to evaluate a paid platform is to ask AI itself. Context matters, even with seemingly simple questions. I explain what I am trying to accomplish, what I expect from the tool, and what my goals are for execution. I then ask it to prompt me for additional details so it can help me think through the decision.

When I do this, I ask for one question at a time. Better context leads to better guidance.

Unlike many human experts, AI has no ego. It does not care which platform you choose. That indifference is an advantage for people like you and me.

AI is a hammer. We get to choose the nails, and how we want to drive them.

Appendix

Prompts for Working With AI as a Collaborative Assistant

The prompts that follow are not recipes or shortcuts.

They are patterns for thinking clearly with AI.

They are designed to slow the exchange, surface assumptions, and keep judgment where it belongs. Replace the bracketed text with your own goals, context, or constraints. Use them selectively. You do not need all of them for every task.

Their purpose is not to make AI more powerful, it is to make your use of AI more deliberate.

Phase 1: Orientation

Establish shared understanding before action

1. Establish Intent First

> I want to accomplish [specific outcome].
> The context for this work is [background, audience, constraints].
> Before offering suggestions, tell me what you understand my intent to be and what might be missing.

2. Invite Clarifying Questions

> I want your help with [task or problem].
> Please ask me clarifying questions, one at a time, until you feel you have sufficient context to
> proceed.

3. Surface Assumptions

For this task, what assumptions are you making about [topic, audience, goals]?
Please list them so I can confirm or correct them before we continue.

Phase 2: Exploration

Examine possibilities without rushing to conclusions.

4. Ask for Alternatives, Not Answers

Given my goal of [goal], what are several different ways to approach this, and what are the tradeoffs of each?

5. Identify What Not to Do

Based on my stated goals, what approaches or actions would likely be ineffective or counterproductive, and why?

6. Invite Uncertainty

Where are you least confident in your guidance on this topic, and what factors contribute to that uncertainty?

Phase 3: Judgment and Fit

Evaluate usefulness, reasoning, and boundaries.

7. Request Reasoning Transparency

> As you respond, explain how you arrived at your conclusions so I can evaluate the reasoning, not just the result.

8. Slow the Exchange

> I'd like to proceed deliberately.
> After each response, pause and ask whether I'd like to continue, adjust direction, or provide more context.

9. Test for Fit, Not Capability

> Given what you know about my goals and constraints, are you the right tool for this task, or would another approach or platform be better suited?

10. Close With Reflection

> Before we finish, summarize what you believe I am trying to accomplish, the key considerations we uncovered, and any open questions I should reflect on.

These prompts work because they do not ask AI to replace judgment.

They ask it to participate in it.

AI does not need to be certain to be useful.

It does not need to be fast to be effective.

Used this way, AI becomes a collaborative assistant rather than a shortcut, helping you think more clearly without taking responsibility away from you.

A Closer Look at DNA Files

What follows are three examples of what is referred to as a DNA file. Each sample will include the prompt, and the resultant DNA file in markdown (.md) format.

1. Personal DNA Example

The prompt:

I would like to create a Personal DNA file that defines my identity, temperament, communication style, strengths, constraints, and values.

This file should guide how you interact with me in future conversations.

Please ask clarifying questions if necessary before generating the file.

Format the output as a clean Markdown document suitable for saving as a .md file.

 # Personal DNA

Name: Jordan Ellis

Identity

I am a mid-career professional transitioning into independent consulting.

I value independence, competence, and long-term thinking.

I prefer thoughtful conversations over rapid back-and-forth exchanges.

Temperament

- Analytical but reflective

- Slow to trust, loyal once committed
- Motivated by mastery rather than recognition
- Energized by solving meaningful problems

Communication Preferences

- Clear, structured responses
- No hype or exaggerated claims
- Avoid slang and overly casual tone
- Bullet points are helpful
- Summaries at the end of complex responses

Strengths

- Pattern recognition
- Strategic thinking
- Calm under pressure
- Willingness to revise decisions when presented with better information

---

Constraints

- Can overthink before acting
- Hesitant with self-promotion
- Dislikes ambiguity without boundaries

Core Values

- Personal responsibility
- Integrity in communication
- Continuous improvement
- Long-term impact over short-term wins

Interaction Guidelines

- Offer thoughtful challenge when needed
- Help clarify tradeoffs
- Do not assume urgency unless stated
- Prioritize depth over speed

2. Business DNA Example

The prompt:

I would like to create a Business DNA file for my company.

This file should define our positioning, tone, audience, decision principles, and boundaries.

It should help guide marketing, messaging, and strategic decisions.

Format it as a Markdown document suitable for long-term reference. Please ask clarifying questions if you need more information. Ask them one at a time so I can respond.

 # Business DNA

Company Name: ClearPath Advisory

Core Positioning

We provide strategic clarity for small businesses experiencing growth complexity.

We are not general consultants.

We specialize in operational alignment.

Audience

- Founder-led companies
- 5–50 employees
- Revenue between $1M–$10M

- Experiencing growing pains, not survival crises

Tone

- Direct but calm
- Analytical, not motivational
- Evidence-based
- Practical over inspirational

What We Do

- Diagnose operational friction
- Clarify priorities
- Create decision frameworks
- Improve internal alignment

What We Do Not Do

- Crisis PR
- Emotional coaching
- Startup incubation
- Social media growth strategy

Decision Principles

- Clarity over complexity
- Long-term viability over quick wins
- Client capability over dependency

Marketing Boundaries

- No urgency language
- No income promises
- No exaggerated testimonials

- No fear-based messaging

Success Metrics

- Client independence increases
- Internal systems become repeatable
- Leadership decisions become faster and calmer

3. Project DNA Example

The prompt:

I am starting a new project and would like to create a Project DNA file.

This file should define the purpose, scope, audience, tone, constraints, and success criteria for this specific project.

It should be focused and practical.

Format it as a Markdown document suitable for saving as a .md file. Please ask clarifying questions if you need more information. Ask them one at a time so I can respond.

Project DNA

Project Name: First-Time Manager Toolkit

Status: Active

Purpose

Create a concise digital guide to help new managers navigate their first 90 days.

This is a practical resource.

It is not a leadership philosophy book.

Audience

- Professionals recently promoted to management
- Managing small teams (3–10 people)
- Limited formal leadership training

Scope
Include:
- Communication basics
- Delegation frameworks
- Feedback structures
- Common early mistakes
Exclude:
- Executive leadership theory
- Organizational psychology deep dives
- Corporate politics strategy

Tone
- Clear and structured
- Calm and supportive
- Practical
- No motivational rhetoric

Constraints
- Under 20,000 words
- Actionable checklists
- Examples in plain language
- No industry jargon

Success Criteria
- Reader feels less anxious
- Reader applies at least one framework immediately
- Reader reports improved clarity within 30 days

Glossary

A

Agent – For purposes of AI, an agent is a process you assign to AI, instructing it to perform tasks such as sorting email or making purchases. AI is acting on your behalf, making decisions and performing tasks per your instructions.

Agentification – A recent, popularized term meaning a process has been handed over to AI to complete per your instructions.

Alignment – when you and AI agree on context, outcomes, and results, you are in alignment.

Artificial General Intelligence (AGI) – A theoretical concept for machines that could actually reason and perform like humans without task-specific training.

Artificial Intelligence (AI) - A phrase coined by John McCarthy in 1955 as part of a proposed program at Dartmouth College titled "*Dartmouth Summer Research Project on Artificial Intelligence.*" The term is now widely, but often imprecisely, used to describe Large Language Model (LLM) platforms.

Automation – Using machines or computers to complete tasks that are typically repetitive and benefit from standardization reducing direct human involvement.

B

Bias –When a person or program consistently favors certain perspectives, patterns, or assumptions, leading to generalized conclusions that may not fully reflect the facts or the broader context.

Black Box – A system whose internal workings are not visible or easily understood. Outputs are produced, but the reasoning behind them remains opaque, making the system feel mysterious or difficult to evaluate.

C

Chain of Thought – A structured progression of reasoning from one idea to the next in service of a specific question or task. Unlike stream of consciousness, which may wander freely, a chain of thought follows an ordered path toward a conclusion.

Context – For purposes of AI, context is the relevant information you provide to define the scope of a task. It includes what matters, what does not, the intended audience, tone, and constraints. Context narrows AI's responses from its broad training to your specific subject and goals.

Context Window – The finite amount of information an AI system can consider at one time. It functions as a viewport into a project's context. When too much information is included, important details can be diluted. When too little is provided, responses may lack accuracy or relevance.

D

Data Training – The process by which an AI system learns patterns, relationships, and relative importance from large collections of data. Training does not give AI understanding or intent, but shapes how strongly different information influences its responses.

Decision Support - The use of AI to assist decision-making by providing analysis, options, or perspective. In a human-centered posture, AI supports human judgment. In a more automated posture, decisions may be delegated to AI within defined constraints.

Determinism – The degree to which an AI system produces the same output when given the same input. Deterministic systems are highly repeatable. Most modern AI systems introduce variability to produce more flexible responses.

E

Emergent Behavior – Patterns of response that reveal themselves through use. Emergent behavior reflects what an AI system has learned and is influenced by context, but context never fully determines behavior.

F

Fine-Tuning – The process of improving the reliability of AI responses by refining how the system is informed and used. In practice, this means adjusting context, prompts, and constraints rather than altering the underlying model.

Foundation Model – A broadly trained AI system designed to support many types of tasks. It provides the underlying capability that users shape through context, prompts, and constraints. Different platforms may use different foundation models optimized for specific types of work, such as text generation, programming, or image creation.

G

Guardrails – The limits applied to an AI interaction through context and instruction to keep it on task and within intended boundaries. Like an employee handbook, guardrails define acceptable behavior without dictating every action.

H

Hallucination – A human term for AI responses that sound convincing but are poorly grounded. AI does not hallucinate. When context or guardrails are insufficient, it fills gaps with language that fits the conversation, not necessarily the facts.

Human-in-the-Loop – You. AI does nothing on its own. Like a textbook or a Magic Eight Ball, it only responds when engaged. Humans provide the necessary context, constraints, and judgment that shape how AI is used and interpreted.

I

Inference – What AI derives from operator input at the time it is used. During inference, the system applies what it has learned from training and user-provided context to generate a response.

J

Judgment – The important ingredient humans bring to the conversation. When AI generates and humans evaluate and judge, the process works.

L

Large Language Model (LLM) – A system often referred to as AI that generates text by predicting what words are most likely to come next, based on patterns learned from large amounts of previously available text.

M

Model Drift – When AI responses gradually move away from the user's intent. This often occurs in long conversations or when context becomes unclear, overloaded, or inconsistent.

Multimodal – A description of AI systems that can work with more than one type of input or output, such as text, images, audio, or video, rather than a single mode alone.

O

Optimization – Adjusting how AI is applied in order to improve the usefulness or consistency of its output for a particular task, such as writing, image generation, or analysis.

Overfitting - When an AI system is shaped too closely around specific examples, reducing its usefulness in new situations. Think of a tuxedo altered so many times it only fits one person, one night.

P

Platform-Agnostic – An approach that avoids loyalty to any one platform, recognizing that no single tool is right for everything.

Predictive Text – A description of how many AI systems generate language by predicting the next word or phrase based on available context.

Prompt – What you give AI to shape its response, including the task, context, and constraints.

Prompt Engineering – The study, art, and practice of shaping prompts so AI responses are more useful, reliable, and aligned with intent.

R

Reasoning Transparency – Asking AI to explain how it arrived at a response so you can judge the reasoning, not just the answer.

S

Self-Correction – Asking AI to revisit its response, check its work, and correct errors in order to favor accuracy over conversational elegance.

Selective Engagement – Choosing when and how much context to bring forward so AI remains focused on what matters in the current task.

Stochastic Output – AI responses influenced by probability rather than fixed rules, allowing small variations in output even when the same input is provided. This controlled variability produces results that are plausible rather than perfectly repeatable.

T

Tool vs Assistant – A way of thinking about how AI is used. When treated as a tool, AI performs a specific task. When treated as an assistant, it helps shape thinking, clarify effort, and guide how a task is approached. The distinction lies not in the system, but in how it is used.

V

Verification - The responsibility of the user who seeks correctness over cleverness. It is the act of checking, confirming, and applying judgment rather than accepting AI output at face value.

Index